THE
CRIMSON GIFT

Previously published under the Title "That Bloody Book"

DAVID P. MCINTYRE

All Bible References are from the New King James Bible (NKJB)

THE CRIMSON GIFT

Bennett books may be ordered through booksellers or by contacting:

Bennett Media and Marketing
1603 Capitol Ave., Suite 310 A233
Cheyenne, WY 82001
www.thebennettmediaandmarketing.com
Phone: 1-307-202-9292

ISBN: 978-1-964296-07-4 (softcover)
ISBN: 978-1-964296-08-1 (eBook)

Printed in the United States of America

Reviewer's Comments

"I have been reading my Bible every day for over forty years. I have been teaching two weekly men's Bible studies for over thirty years. I knew the thread of the symbol of blood was very significant from Genesis to the crucifixion. Never until David's research and this book did I know, however, just how incredible the chemical makeup of blood is to our salvation story. Absolutely amazing book! "

Dr. Bob Barnes
CEO Sheridan House Family Ministries

"Wow! Thanks for bringing such amazing clarity to one of the most misunderstood parts of the Bible."

Rick Weber
President Sheridan House Family Ministries

INTRODUCTION

Throughout the Bible is a crimson thread that begins in Genesis and continues to Revelation. It starts subtly with Adam and Eve, is expounded more in the books of Moses, is there behind the scenes in the rest of the Old Testament and is fully revealed with Jesus on the cross. It is the Crimson Gift, the blood of Jesus.

In this book, I would like to take a close look at one of the most common themes throughout the Bible, Blood. The Bible has often been characterized as a very bloody book. Some assume that it is full of stories of bloody conflict, savage cruelty, and inhumanity. Wars have been fought based on interpretations of what is in its pages. Others have been severely persecuted for simply believing in what it says. Let's see what it does say.

I found 43 out of the 66 books of the Bible mention blood. The one central theme that seems to pervade the whole Bible is the idea that blood must be shed for sin.

> *For the life of the flesh is in the blood, and I have given it to you upon the altar to make atonement for your souls; for it is the blood that makes atonement for the soul. (Lev 17:11)*

That statement is an incredibly accurate scientific statement especially when we consider that it was written by Moses over 3400 years ago. That was a long time before modern medicine. In fact, even into the late 18th century doctors were bloodletting to cure disease. Doctors removed 3.75 liters of blood, for example, from our first president George Washington to cure a sore throat.

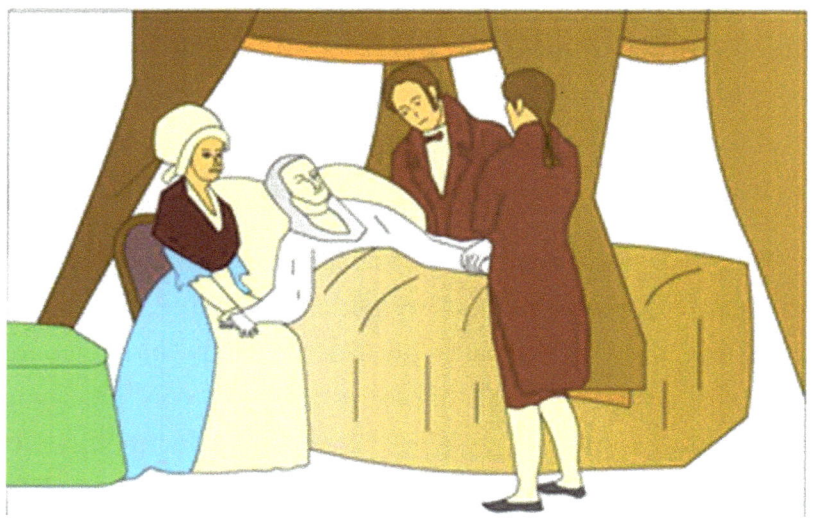

He died soon afterward. Those same doctors could have learned a valuable lesson if they had read and understood the message in Leviticus 17:11. The life of the flesh is in the blood. You can live without an arm, a leg, a kidney, a lung, and many other parts of the body. If you lose enough blood, though, you die even if everything else is working perfectly. The life of the flesh is in the blood.

My wife as a registered nurse saw the truth of Leviticus 17:11 many times in the hospital. Many years ago, she worked in the ER when a young man was brought in the victim of a motorcycle accident. As they worked on him on the table, he looked over at my wife and asked her not to call his wife because she would only worry. He died a half hour later.

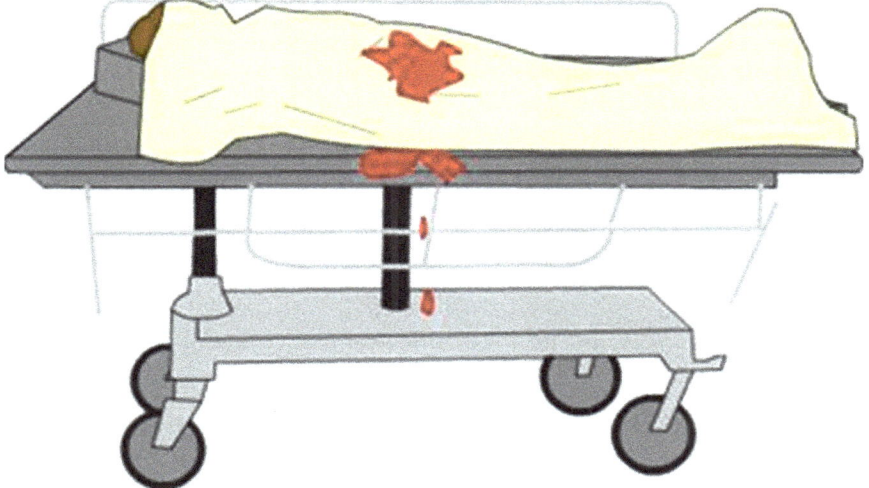

His heart was OK. His lungs were OK. His mind was in tack. They just could not stop the bleeding. When there was not enough blood, he died because the life of the flesh is in the blood.

The blood theme and its sacrifice for sin is seen all the way from Genesis to Revelation. God builds the concept slowly, carefully teaching us and expanding on the concept as we read through the Scriptures. We are going to explore the theme together following God's lead through the passages in the next chapter. The chapter after that examines the more difficult blood concepts. I believe that to really understand what God is saying, you must study this marvelous liquid called blood. We will spend a few chapters on just what blood does in the body. Then we will apply what we learn. I believe that God has given us the best analogy in blood in the physical realm to teach us what He is trying to accomplish in the spiritual realm. Are you ready for the challenge?

BLOOD SACRIFICE

O ne central theme of the Bible is that blood is required for sin. We are introduced to the concept early in Genesis. God creates the heavens and the earth in Genesis 1. He then creates Adam and makes him His manager of His creation and creates a helpmate for him, Eve, in Genesis 2. God gives every tree in the garden for food except one, the tree of the knowledge of good and evil. He asks that Adam (Gen 2:17) not eat of that tree and warns him that if he does, he will die. Satan enters the scene in chapter 3 in the form of a serpent, beguiles Eve, and she eats of the fruit. Adam follows suit. Their actions bring sin into God's creation. God shows them how serious the sin is by cursing the serpent, then the rest of creation, and tells Adam that it will now be much harder for him to live on the land. Then God seems to add a footnote in Genesis 3:21:

"Also, for Adam and his wife the LORD God made tunics of skin and clothed them."

There is only one way to obtain skins for tunics; an animal must die. The first blood is now shed in the Bible. When we read those early chapters in Genesis, it is easy to gloss over this inconspicuous verse. But wow! God does not view sin the way we do. It is serious business with Him. We like Eve quickly try to justify our actions by blaming something or someone else. Eve blamed the serpent. Adam blamed Eve. Everyone says the "devil made me do it". God puts the blame squarely on our shoulders. We make the decision to sin. We also do not see the consequences of our sin, but God does. To us it is just a little infraction. What does it matter if we taste a little of the fruit. God sees the one disobedient act as leading to others and then to even more acts. He views the sin like a physician might view the survival of a single cancer cell in the body. If that cancer cell survives, it becomes 2, then 4, then 8 and eventually becomes a mass that kills the body. For the good of the patient, the physician must do everything he can to destroy all the cancer cells. In the same way God sees where sin leads from the very beginning. He saw all the subsequent disobedient acts of mankind and how they all lead to other sins. Eventually it would lead to the destruction of mankind and His creation. He could not just sweep it under the table. He had to do something about it. He pronounced the curse and bans Adam and Eve from the Garden, so they cannot eat of the tree of life and live forever (Gen 3:22). He had to limit the destructive power of sin. At the same time, though, He hints at redemption. Blood was shed to cover Adam and Eve's sin. The curtain is raised a little to reveal more of God's plan to save man.

Turn the page to the next chapter and God teaches us that blood is the only acceptable sacrifice for sin. The scene is the story of Cain and Abel each bringing their sacrifices to God. Cain was a farmer tilling the ground. Abel raised sheep. Each brought their harvest to the Lord. Cain brought fruit (Gen 4:3). Abel brought the first born of his flock.

Cain and Abel

Abel's offering was accepted (Gen 4:4) but Cain's was not (Gen 4:5) and it made Cain very angry. We may have felt the same way as Cain after all he brought the first portion of all his increase. God, though, had been very specific that only shed blood can make an atonement for sin. Cain should have traded the fruit for a sheep and made his offering. He didn't. He tried to do it his way. God is not obligated to accept our way. He created the heavens and the earth. It is His. Heaven is His. Entrance into His heaven is only by the door He opens for us. Leviticus 17:11 tells us that it is only through shed blood. Cain should have known better.

God raises the curtain on the blood theme in Genesis 8. Noah and his family have spent over a year on an ark while God brought a devastating flood that destroyed all animals and mankind outside the ark's protection. God calms the seas, stops the rain, and provides paths for the waters to recede off the earth. Noah finally opens the ark and the animals inside exit. In Genesis 8:20, Noah builds an altar and sacrifices every clean animal and bird on it. Blood is again shed for the remission of sin. This time it is for himself, his wife, his sons, and their wives.

The earth is repopulated through Noah's sons and further on in Genesis we meet Abraham. He is told to leave Ur of the Chaldeans and go to the land of Canaan. He eventually obeys. Later, God promises a son to him and his wife Sarah. His descendants would number like the stars. Abraham and Sarah were older, and Sarah was beyond childbearing years. It did not occur right away. In fact, it took another 25 years before Isaac was born. Isaac grows up and becomes a teenager and then Abraham is tested in Genesis 22. He is asked to take his son up to a mountain in Moriah and sacrifice him there. The son he waited 25 years to see born, has loved for a dozen years, or more must now be sacrificed. What a test of Abraham's faith! Abraham's faith was strong, though, and he took his son to the mountain and placed him on the altar. Just as he is about to plunge in the knife, an angel of the Lord stops him and directs him to sacrifice a ram caught in the thicket instead.

God lifts the curtain a bit higher. Blood must be shed for sin, but it can be a substitute blood, in this case a ram. Many centuries later another son would be sacrificed on this same spot to provide lifegiving blood as the ultimate sacrifice.

The blood theme continues into Exodus. Moses has been raised up to lead the children of Israel, Abraham's descendants, out of bondage in Egypt. It is the last night before they are to leave, he is told in Exodus 12 to have every family kill a firstborn male lamb or goat, one for each family, eat it, and place the blood on the doorposts and lintel of the house. Can you not see the cross on the door?

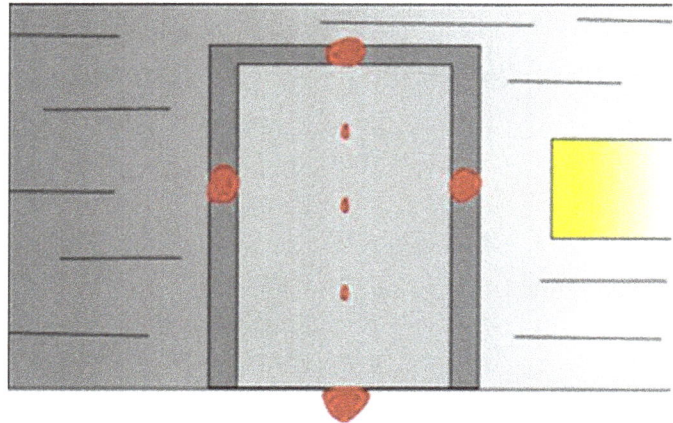

God then brings in the death angel that kills all the first born of the Egyptians but spares all those Of Israel who have obeyed. Israel has celebrated this "passing over" event on the 10th day of the first month of the Jewish calendar ever since. We now have a little more information. The blood will be substitutionary blood from a firstborn male with no blemish, a lamb for a household, and we must partake of it.

Further on in Leviticus chapter 16:15-16, we see instructions for the high priest to bring the blood of the sacrificial lamb into the Holy of Holies and sprinkle it on the mercy seat. The lamb is now being sacrificed to atone for the sins of the nation. It has gone from a lamb for an individual, to a lamb for a household, to a lamb for a nation. There is also an interesting passage in the same chapter Leviticus 16:20-22. After the first goat is sacrificed, a second live goat is brought in. The priest was to lay both his hands on the live goat, confess all the sins of Israel on it and send it off into the wilderness.

The sacrifice was not only to be an atonement for sin but would take the sin away.

As far as the east is from the west, so far has He removed our transgressions from us. (Psalm 103:12)

The atoning sacrifice was practiced throughout the Old Testament all the way into the new. John the Baptist sees Jesus coming toward him in John 1:29 and exclaims:

Behold the Lamb of God who takes away the sin of the world.

We have gone from a lamb for a man to a lamb for a household, to a lamb for a nation to a lamb for the world. Jesus is the ultimate blood sacrifice. It is through His shed blood that we obtain forgiveness for sin.

> *In Him we have redemption through His blood, the forgiveness of sins, according to the riches of His grace. (Eph 1:7)*

> *But now in Christ Jesus you who once were far off have been brought near by the blood of Christ. (Eph 2:13)*

The life of the flesh is in the blood. It is the blood of Jesus that gives real life, real forgiveness of sin. The story does not end there, however, there is more.

JESUS'S BLOOD

The atoning sacrificial blood of Jesus is to be far more than just a one-time sacrifice for the sins of the world. The trail does not stop with John the Baptist's introduction of Jesus as the Lamb of God. It continues. John 6 verses 53 through 56 says:

> *Then Jesus said to them, "Most assuredly, I say to you, unless you eat the flesh of the Son of Man and drink His blood, you have no life in you. Whoever eats My flesh and drinks My blood has eternal life, and I will raise him up at the last day. For my flesh is food indeed and My blood is drink indeed. He who eats My flesh and drinks My blood abides in Me and I in him.*

Whoa, this is really getting weird now. If you ate just before reading this, your stomach may be doing flipflops right about now. It's OK if you want to take a break. It is going to get a little bit worse, but you must see it all the way to fully understand what God is saying.

I Peter 2 verse 24 says:

> *Who Himself bore our sins in His own body on the tree, that we having died to sins, might live for righteousness by whose stripes you were healed.*

The verse echoes another verse in the Old Testament, Isaiah 53:5:

But He was wounded for our transgressions, He was bruised for our iniquities; The chastisement for our peace was upon Him, and by His stripes we are healed.

Somehow this blood shed by Jesus on the cross has healing power. Am I not only supposed to eat and drink the blood but put it on my wounds to heal them? There is still more though. I John 1: 7 says:

But if we walk in the light as He is in the light, we have fellowship with one another, and the blood of Jesus Christ His Son cleanses us from all sin.

Wow, now I am to wash with it. The blood message continues all the way into Revelation where chapter 12 verses 10 and 11 say:

Then I heard a loud voice saying in heaven, "Now salvation, and strength, and the kingdom of God, and the power of His Christ have come, for the accuser of our brethren, who accused them before God day and night, has been cast down. And they overcame him by the blood of the Lamb and by the word of their testimony, and they did not love their lives to the death.

The image that I conjure up in my head at this point is a bit R or even X rated. I imagine a man with blood dripping from his mouth, wounds covered in blood, arms and face washed with blood, and the man throwing blood balls at the devil. The Bible really is a bloody book.

If you want to make sense and understand these images you must study blood. We already know one thing; the Bible tells us that the life of the flesh is in the blood (Lev 17:11). I believe that when we fully comprehend what blood does and all its properties, the Bible verses will become clear. Stay with me as we examine blood in the next chapters.

DAVID P. MCINTYRE

BLOOD PROPERTIES

Over the years I have had the privilege of working for a Medical Instrument Company (Beckman Coulter) as a senior technical instructor. The company produced several blood analysis instruments and reagents used to diagnose diseases by examining blood. My students attended class to learn how to operate, maintain and troubleshoot the systems. Most were medical technologists, but many were medical doctors and research Ph.Ds. I taught them the system operation and they in turn taught me about blood. They gave me a great deal of insight on what blood does, how testing helps them diagnose and treat disease, blood properties, and I had an inside look at some of the latest research findings on this precious liquid we call blood.

They test blood because it goes everywhere in the body. If anything is wrong internally in the body, evidence of the event is often dumped into and carried along in the blood. I had a personal example of that many years ago, when I woke up at 3 AM with some chest discomfort. I thought it was indigestion, took something for it, it subsided, and I went back to sleep. Later I got up and went to work but I felt a bit off and I asked my boss if I could go home. I waited a couple of hours while they found a substitute instructor and then drove home. After consulting my wife, I called my doctor and drove over to see him. At that point, he read me the riot act and called an ambulance to take me to emergency. He was thinking I had a heart attack. They examined me in the emergency room. My blood pressure was normal, EKG was normal, and I had no pain. After several hours, they were about to send me home when the blood work results came back. The heart enzymes were high, and they knew it was a heart attack. There was a block off the major artery. Two stints were inserted to keep the blocked arteries open. In a couple of days, I was back at work none the worst for the experience. The blood test told the story.

When blood is drawn into a tube and placed in a test tube rack, it will settle out into two distinct fractions. The top liquid fraction is called plasma (or serum depending on what is already in the tube). The heavier red bottom fraction contains cells, red cells, platelets, and white cells. I believe that blood provides the best analogy God could ever devise to illustrate what He is trying to do for us spiritually. To understand the analogy, however, we must study this phenomenal liquid called blood and see what it does in the body. It was then I understood why "The life of the flesh is in the blood".

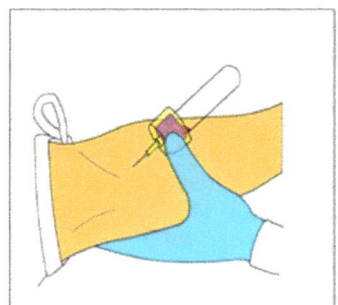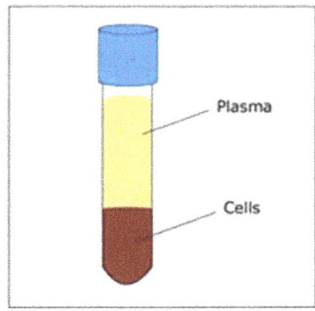

Plasma

The cells in the body other than those in the blood are largely locked into place. They may stretch or shrink but they cannot move. Muscle cells, skin cells, lung cells, heart cells, etc. must stay in place on the job so that the whole body can function properly. They may die and be replaced but they cannot take a vacation. They cannot go out shopping for any needed supplies. Anything those cells need to survive must be brought to them. The liquid portion of the blood supplies several of the chemicals needed for survival.

Food

Every living thing must have food to survive. The individual cells are no different. Dissolved in the liquid portion of the blood are proteins, sugars, cholesterol, and triglycerides the cells need to live and function. The cells cannot leave and shop for these items, so the "store" comes to them. The cells can then extract exactly what they need when they need it. If the blood did not supply these items, the cells would literally starve to death. The life of the flesh is in the blood. But that is not all.

Water

Living things cannot survive without water and neither can the individual cells. Either too little or too much will kill them. They will either dry up or blow up. The foundational component of blood is water. When the cells are surrounded by blood, they can take in water from the blood if they do not have enough or get rid of excess water if they have too much. The water in the blood maintains the proper amount of water in the cells so they can survive and function as needed. That is also why we need to drink enough water so that the blood carries the right proportion of water. There is more.

Trace Materials

Certain chemical processes may require some trace materials such as magnesium, calcium, iron, zinc, copper, as well as vitamins to support the chemical reactions within the cells. Without these materials the cells may not be able to produce the materials needed to repair themselves, replace themselves or function as the cell is designed. Then the whole body suffers. If you are a cook creating those fantastic dishes for family and friends, you probably have a whole arsenal of herbs and spices to add to the food to enhance the taste. You do not use all of them all the time, but those special dishes just are not the same if any herbs or spices are missing. It is like that with the cells. They may not need every mineral or vitamin for every reaction but when they are needed, it is important that item be available. If muscle cells do not have enough magnesium, for example, the cells may not be able to contract or expand properly upon command. The result is painful muscle cramps. The cells do not have a pantry where they can obtain the materials they need. Their pantry is the blood surrounding them. If the blood does not have it, they are out of luck. The life of the flesh is in the blood. There is still more.

Enzymes

Certain chemical reactions require an activator to start the reaction or a chemical to sustain the reaction or something to guide the reaction to the desired outcome. These activators, sustainers, and guides fall into a class of chemicals called enzymes. If they are missing, there may be no reaction or the wrong reaction leading to the wrong chemical product. We may be more familiar with enzymes needed for digestion, but individual cells also need some of these enzymes on a smaller scale for some of their chemical processes as well. Again, the cells do not have a storage area to supply the needed enzymes. They must obtain what they need from the surrounding blood. If there are no enzymes, there is no or the wrong reaction product. The life of the reaction is dependent on the blood. That is not all.

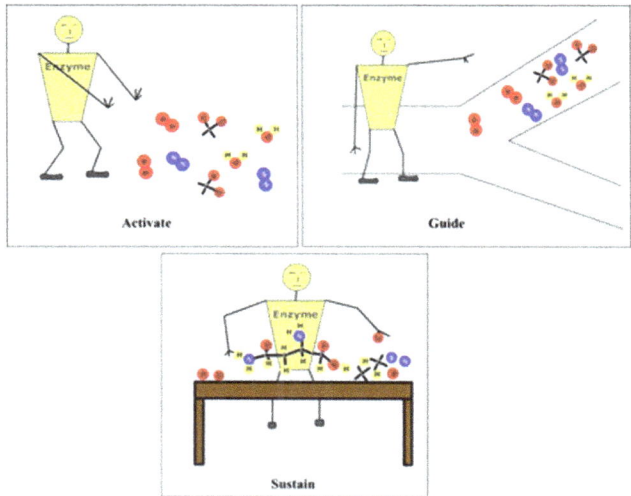

Waste Removal

Every chemical reaction produces waste byproducts. These in turn must be removed from the cell. Imagine what would happen if you did not have garbage pickup for weeks or months on end. The waste pileup would be enormous. The smell would be horrible, and the possibility of disease would be unbearable. Your home would become unlivable. The cell is no different. The waste products of cell chemical reactions must be removed. If they are not, the environment within the cell will deteriorate and the cell will eventually die.

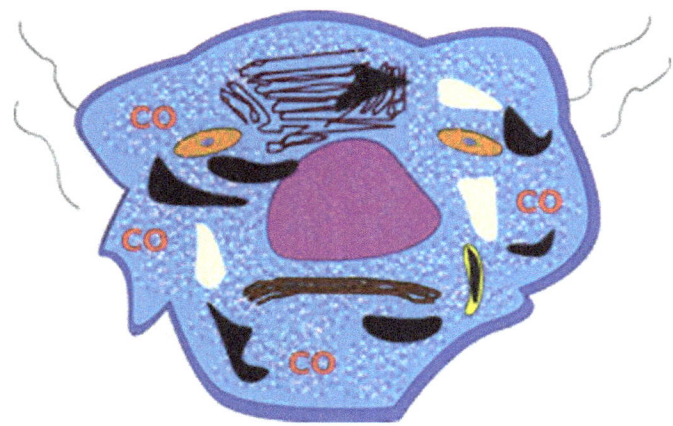

The garbage removal service for the cell is the blood itself. It is interesting that the same blood that brings the food to the cell also removes the waste from the cell. The cell does not mix it up. Somehow the portals in the cell membrane open to take in food while other portals open to remove the waste. Again, the cells are saved by the blood. The life of the flesh is in the blood. That is still not all.

Antibodies and Antitoxins

Sometimes the cells may encounter chemicals or other cells that may do them harm. They may come in externally or somehow the blood itself may be infected. The cells locked in place may have little defense against chemical or cellular attacks. Again, the blood itself can come to the rescue with antitoxins to neutralize the effects of some chemicals or antibodies that target unwanted cells for destruction.

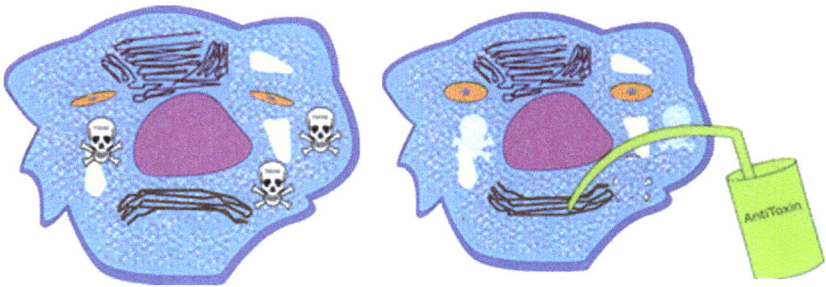

We will explain more about antibodies and cell interaction when we cover the cellular components of the blood. The blood then supplies a measure of protection to the cells everywhere in the body. The blood not only sustains life but protects it as well. There is still more.

Salts

Many chemical reactions can change the surrounding liquid to a lower or higher PH value. In short, the liquid can become more acidic or basic. Strong acids burn and can leave scars on the skin. Strong bases dissolve biologic material including flesh. Individual cells within the body are no different. They can be burned if the surrounding liquid becomes too acidic or dissolved if it becomes too basic. The very reactions in the cell can change the surrounding PH to levels that may be dangerous to the cell. Again, blood to the rescue. Dissolved in the liquid portion of the blood are salts. They serve to neutralize the acid or base and bring the PH back to a safe level. The safest place for the body's cells is surrounded by blood.

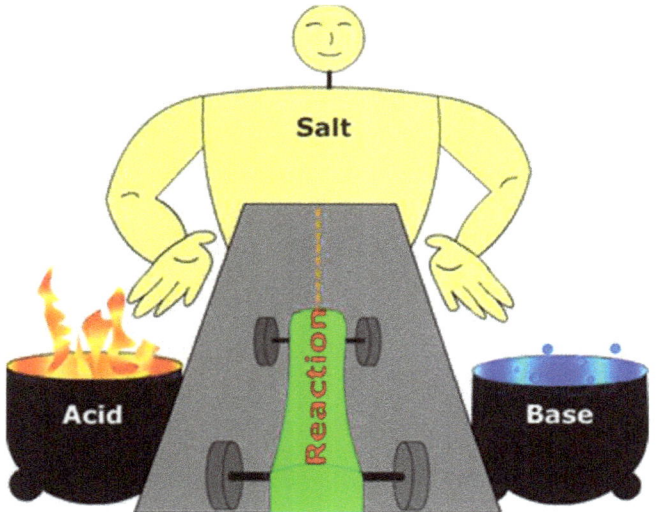

Cells

The liquid portion of the blood is vital to the survival of the individual cells throughout the body. Cellular components are also extremely important as well. There are three types of cells in the clump at the bottom of a blood tube:

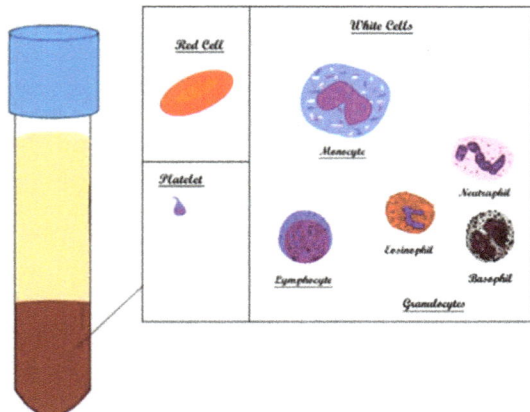

Red Cells

Oxygen is another vital component in body cell survival. Brain cells, muscle cells, tissue cells, etc. all need this vital element to function properly. Red blood cells (RBCs) contain a chemical called Hemoglobin which serves as a carrier for oxygen (O_2) and carbon dioxide (CO_2) molecules. When these cells reach the lungs, oxygen is transferred to the hemoglobin and the cells turn red giving blood its red color. These cells then travel through the arteries to capillaries and then cradle the individual cells throughout the body. The oxygen is exchanged for the waste gas carbon dioxide. At this point the cells turn blue. The RBCs then continue back to the lungs where the carbon dioxide is exchanged for a fresh load of oxygen and the cycle repeats. The RBCs have a biconcave shape and are extremely flexible. The flexibility allows them to squeeze through the smallest capillary hole to reach the cells starving for oxygen.

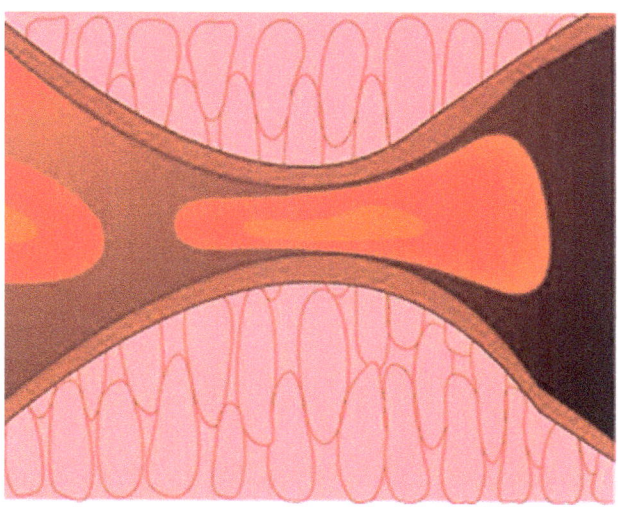

The shape allows them to cradle the oxygen starved cells providing the maximum surface area for efficient transfer of oxygen.

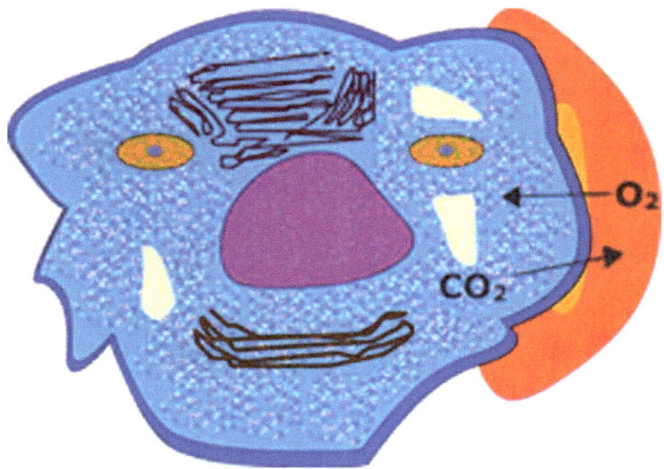

Red blood cells last about 100 to 120 days in the blood and then are recycled. There are approximately 5 to 6 million of them in a microliter of blood and they comprise the largest share of the total cell volume. These cells are probably the most vital of all the cells because of the dire need of oxygen by other cells in the body.

Platelets

Platelets are the second most plentiful cells in the blood with about 150 to 400 thousand per microliter. They are tiny cells, much smaller than red blood cells and they are largely dormant until needed. They go to work whenever the body bleeds either internally or externally. When a bleed occurs, the platelets are activated, they become sticky and bond together. Fibrin is also created to link the platelets together to create a patch over the wound to stop the bleeding. The body then repairs the underlying tissue that caused the bleed. Once the tissue is repaired the patch is carefully removed.

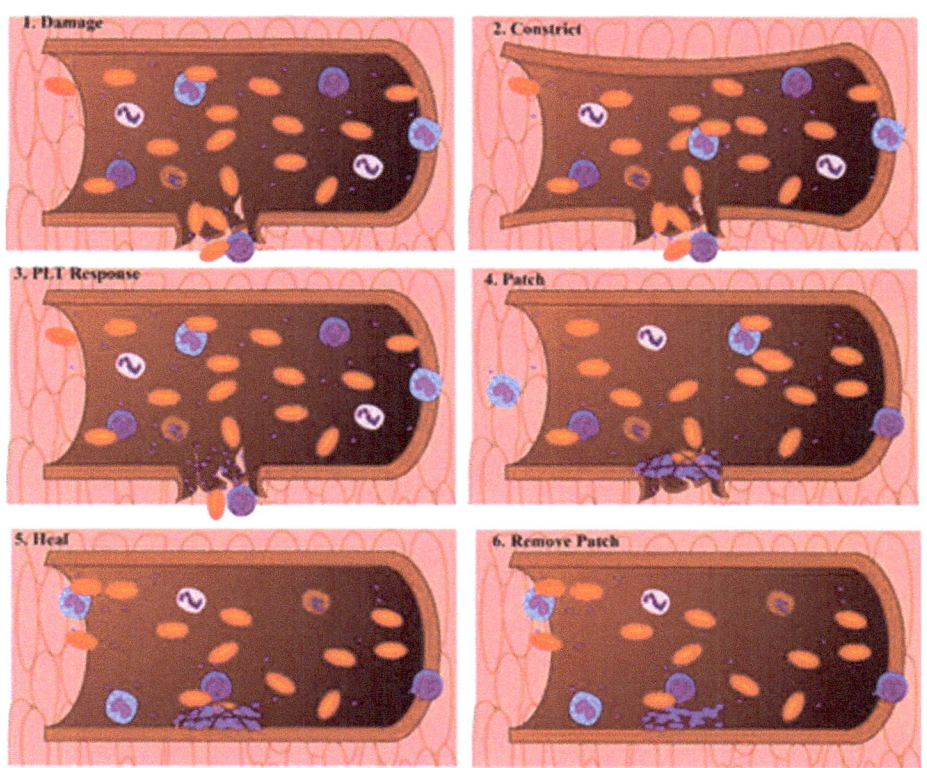

Clotting

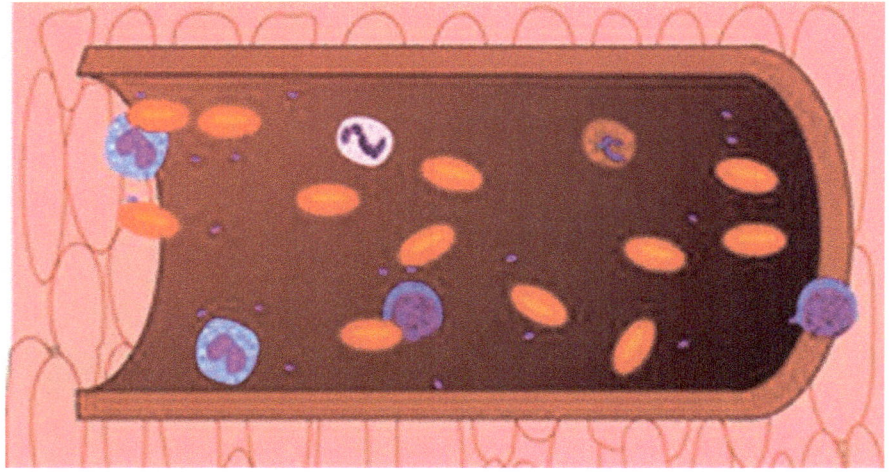

Healthy tissue

This description, of course, is way over simplified. The signaling process involved is extremely complex. The blood must respond properly to the bleed but not too much. If the patch is too big, the blood flow is blocked to vital areas such as the heart or brain and the body suffers a heart attack or stroke or something else. If the response is insufficient, the bleed continues creating other problems or even death. Once the underlying tissue has been repaired, the patch or clot must be carefully removed, usually in smaller pieces. If it breaks off as a complete piece, it blocks somewhere else creating even more problems. The whole process is referred to as hemostasis.

White Cells

The White blood cells (WBCs) are the third cell type in the blood. There are about 4.5 to 10 thousand of them per microliter of blood. These are the soldiers that fight disease invading the body referred to as the immune system. There are three basic types of white blood cells, Lymphocytes, Monocytes, and Granulocytes. The interaction between these cells is very complex. Many scientists much smarter than me have studied the interaction for years. I am going to greatly oversimplify the interaction just for the sake of clarity.

It seems that all the body's cells have a mechanism attached to the surface of the cell so that the white blood cells can identify them as belonging to the body and not foreign to it. Just as every company has identification badges to identify employees, the cells in the body carry personal identification to identify them as belonging to the body.

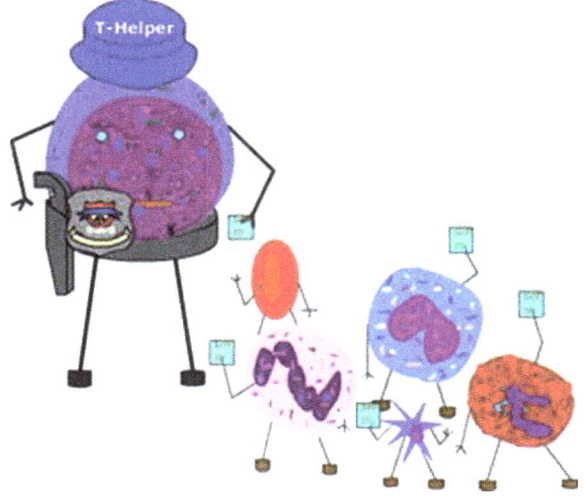

Now imagine a foreign cell invades the body. Just as a company has guards that check all visitors, a subset of the Lymphocytes called T- Helper cells are the guards on duty to check for invading cells. The invading cell may be bacteria, virus or just a cell without the proper identification (non-self).

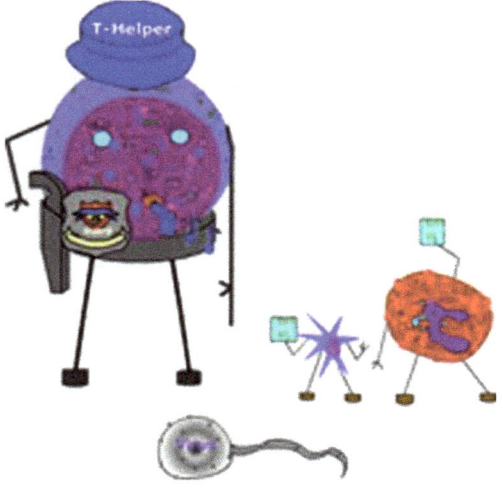

The T-Helper cell now mounts a response. In some cases, a lymphocyte called a natural killer cell takes out the invader immediately.

In another case, the response may be to activate a Lymphocyte known as a B-cell to produce antibodies that will seek out and bind to proteins on the surface of the invading cell.

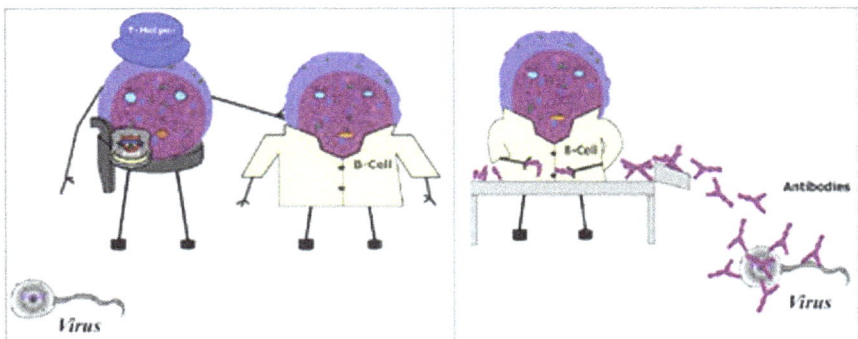

These antibodies mark the cell. The binding is very specific. One end of the antibody matches exactly a protein found on the invading cell. Scientists often describe the binding as analogous to a key and a lock. A key is shaped to open a specific lock not every lock. The antibody binds to a specific protein, not every protein. Specific B-cells produce specific antibodies. Interestingly, our bodies from birth have all those B-cells ready to produce antibodies against foreign invaders before there is any invasion.

The monocyte and granulocytic white cells are continually looking for these marked cells. The other end of the antibody triggers mechanisms to attack and destroy the invading cell.

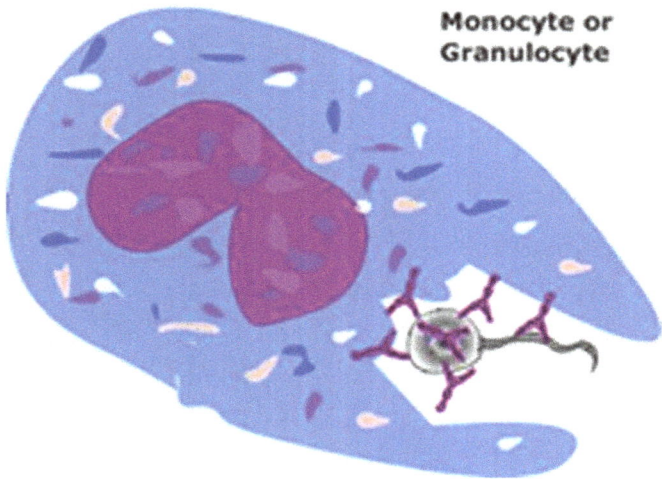

Monocyte or Granulocyte

When the invading cells have all been destroyed, the B-cell antibody production needs to be shut down or at least curtailed. A second Lymphocyte called a T-suppressor cell watches the invading cells and signals the B-cell to stop or limit antibody production.

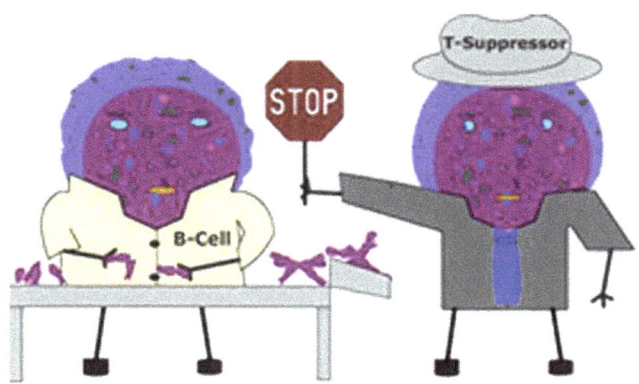

Some antibody production may be maintained in case the foreign cell invades again. Immunization shots purposely inject compromised disease cells (such as viruses and bacteria) to initiate a response and build up antibodies against the disease. The white cells are continuously on duty. Occasionally, however, they can be overwhelmed and that is when we come down with a cold or flu. Once the white cells have mustered enough forces to overcome the invasion, health improves. Some invading bacteria and viruses can be too much for the immune system. E-bola, for example, may kill before the immune system can conquer the invasion. The HIV virus is particularly bad. This virus attacks the T-Helper cells. These cells are the guards on duty. Take them out and there are no guards checking for invading cells and now viruses and bacteria that would normally be taken out by the immune system are free to take over the body. The HIV positive person eventually may die from the common cold. The only real protection against these types of bacteria and viruses is to avoid coming in contact with them in the first place.

THE BLOOD ANALOGY

Blood is the perfect analogy to God's plan for man's redemption. It begins with a personal acceptance of the gift of eternal life.

> *For God so loved the world that He gave His only begotten Son, that whoever believes in Him should not perish but have everlasting life. (John 3:16)*

Once we have accepted the gift, believers are part of the body of Christ. We are the spiritual cells in the ear, the hand, the big toe, and everywhere else in His body.

> *Now you are the body of Christ, and members individually. (I Cor 12:27)*
>
> *So, we being many, are one body in Christ, and individually members of one another. (Rom 12:5)*

Like the cells in any body, we are totally dependent on the blood of Christ to provide everything that we need. We can do nothing by ourselves. Just as the physical blood supplies all the needs of the individual cells in the body, Christ's blood supplies all our spiritual needs.

The Blood nourishes.

Without His flesh and blood there is no spiritual life. Just as the physical cells must allow food to enter in from the blood, we must eat and drink from the blood of Christ. Remember the verse that we are to eat and drink His blood:

> Then Jesus said to them, "most assuredly, I say to you, unless you eat the flesh of the Son of Man and drink His blood, you have no life in you." (John 6:53)

God is speaking spiritually not physically.

> It is the Spirit who gives life; the flesh profits nothing. The words that I speak to you are spirit, and they are life. (John 6:63)

We like the physical cells must take in the nourishment from the blood, we must partake of the spiritual nourishment from the blood of Christ. If the physical cells do not eat, they die. If we do not eat spiritually, we die spiritually. We also must, like the physical cells, eat on a continual basis. We cannot do it once and survive. We must feed regularly.

Not only does the blood supply food, it supplies water. The blood of Christ supplies living spiritual water.

> Jesus answered and said to her, "If you knew the gift of God, and who it is who says to you, 'Give me to drink', you would have asked Him, and He would have given you living water." (John 4:10)

> Jesus answered and said to her "Whoever drinks of this water will thirst again, but whoever drinks of the water that I shall give him will never thirst. But the water that I shall give him will become in him a fountain of water springing up into everlasting life." (John 4:13,14)

Did you catch the message in the second set of verses? If we drink once, we will never be thirsty again. I believe the Holy Spirit comes in us and continually supplies spiritual guidance. We need to be quiet and listen, but He is always there.

> *And I will pray the Father and He will give you another Helper that may abide with you forever the Spirit of truth whom the world cannot receive, because it neither sees Him nor knows Him; but you know Him, for He dwells with you and will be in you. (John 14:16,17)*

Like the enzymes in the physical blood that activate, guide, and sustain physical reactions in the body, The Holy Spirit is continually working. He is our teacher, comforter, guide. He nudges us in the way we should go or act (activation). He guides us through our problems. He sustains us through our difficulties.

> *But you shall receive power when the Holy Spirit has come upon you; and you shall be witnesses to Me in Jerusalem, and in all Judea and Samaria, and to the end of the earth. (Acts 1:8)*

> *But the Helper, the Holy Spirit, whom the Father will send in My name, He will teach you all things, and bring to your remembrance all things that I said unto you. (John 14:26)*

> *Now when they bring you to the synagogues and magistrates and authorities, do not worry about how or what you should answer or what you should say, for*

the Holy Spirit will teach you in that very hour what you ought to say. (Luke 12:11,12)

The world of course is trying to feed us bad information and lead us in the wrong direction. Satan is behind it all. If we feed on it though, we are feeding on bad blood. Bad blood is anemic. It cannot meet all our spiritual needs. Just as anemic physical blood causes all sorts of physical problems, feeding on the wrong things from the world leads to despondency, fear, worry, etc. We need to be guided, strengthened, and sustained by the Holy Spirit (the good blood) and He can keep us spiritually healthy.

I can do all things through Christ who strengthens me. (Phil 4:13)

Sometimes people claiming to be very spiritual feed on the world and then try to force their way onto everyone else. They even try to prevent the Holy Spirit from doing His job. Physical blood can be blocked by cells from reaching other cells. Such action can lead to serious physical problems like a heart attack or a loss of a limb. Weak churches or Christians can also stand in the way of reaching the world for Christ.

Just as the salts in physical blood can neutralize the harmful effects of chemical reactions and anti-toxins can neutralize harmful toxins, the Blood of Christ can neutralize life's storms. The result is peace.

Peace I leave with you, My peace I give to you; not as the world gives do I give to you. Let not your heart be troubled neither let it be afraid. (John 14:27)

Continually Cleansing

The physical blood is also the garbage collector in the body. The individual cells cannot remove their own waste. If the blood does not take it away, the cells will smother in their own garbage. Just as physical blood removes waste, Christ's Blood removes our sins.

As far as the east is from the west so far has He removed our transgressions from us. (Psalm 103:12)

The Blood of Christ is continually doing it just like the physical blood. There is continual cleansing.

If we confess our sins, He is faithful and just to forgive us our sins and to cleanse us from all unrighteousness. (I John 1:9)

Blood Heals

Blood is also continually involved in the healing process. Just as the platelets patch us up when we are wounded, the Blood of Christ patches up our spiritual wounds. That same blood brings healing to our "spiritual tissue" just as blood brings all that is needed to heal physically. The blood stands as a barrier to prevent further damage to cells from whatever caused the damage in the first place. The Blood of Christ stands as a barrier against the destructive effects of sin to bring us spiritual healing, but it was costly. Jesus had to take on our sin and punishment.

But He was wounded for our transgressions, He was bruised for our iniquities; the chastisement for our peace was upon Him, and by His stripes we are healed. (Isaiah 53:5)

Who Himself bore our sins in His own body on the tree, that we, having died to sins, might live for righteousness – by whose stripes you were healed. (I Peter 2:24)

Blood Protects

Blood also protects the body fighting against disease and invading cells that may harm us. Remember the white cell army. Christ' blood also protects us giving us all we need to fight off the attacks of Satan.

You are of God, little children, and have overcome them, because He who is in you is greater than he who is in the world. (I John 4:4)

We must get behind the blood army and allow it to defend us. We should never try to defend ourselves in our own power. Blood has all the resources to fight disease invasions. There are white cells preprogrammed from birth to produce antibodies against any invader. They just need to be activated by the white cell guards on duty. Natural killer cells at times jump in and kill invading viruses and bacteria. We need the protection the blood provides. We also need the full protection, the full armor, the Blood of Christ provides.

Finally, my brethren, be strong in the Lord and in the power of His might. Put on the whole armor of God, that you may be able to stand against the wiles of the devil. For we do not wrestle against flesh and blood, but against principalities, against powers, against the rulers of darkness of this age, against spiritual hosts of wickedness in heavenly places. Therefore, take up the whole armor of God, that you may be able to withstand in the evil day, and having done all, to stand. Stand therefore having girded your waist with truth, having put on the breastplate of righteousness, and having shod your feet with the preparation of the gospel of peace; above all, taking the shield of faith with which, you will be able to quench all the fiery darts of the wicked one. And take the helmet of salvation, and the sword of the Spirit, which is the word of God; praying always with all prayer and supplication in the Spirit, being watchful to this end with all perseverance and supplication for all saints. (Eph 6:10-18)

In the physical realm, if we overload the immune system by overexposing our body to disease and invading organisms, we get sick and may even die. The same is true in the spiritual realm. Continuing in sin leads to spiritual sickness and death.

Then when desire has conceived, it gives birth to sin; and sin, when it is full-grown, brings forth death. (James 1:15)

The solution is the same both physically and spiritually, get away from whatever is overloading our physical or spiritual immune system.

No temptation has overtaken you except such as is common to man; but God is faithful, who will not allow you to be tempted beyond what you are able, but with the temptation will also make the way of escape, that you may be able to bear it. Therefore, my beloved, flee from idolatry. (I Cor 10:13,14)

Blood Provides All the Needs

The blood nourishes, cleans, heals, and protects. It supplies all the needs of the cell. A cell would be foolish to try to survive apart from the life-giving attributes of the blood. In the same way, the blood of Christ supplies all our spiritual needs as part of the body of Christ. It nourishes, it cleans, it heals, it protects.

And my God shall supply all your need according to His riches in glory by Christ Jesus. (Phil 4:19)

The neat thing too is that the blood never asks the individual cells to come to the source of food, water, oxygen, waste removal site etc. It comes to the cells. The lungs never say to the cells: "When you are in the area, I will give you a fresh load of oxygen and take away your carbon dioxide." No, the red blood cells deliver the oxygen and take away the carbon dioxide directly to each cell. The kidneys never say to the cells: "When you are in the neighborhood, I will take that load of waste off your hands." No, the blood comes to the cells, collects the waste, and takes it away. The stomach and intestines never say to the cells: "Come around by my place and I will give you some food." No, the food is brought to the cells. The immune system doesn't congregate on a base somewhere in the body waiting to be mobilized. No, it is available to the cells 24/7 continuously protecting each cell on site. In the same way, Christ knowing we were locked into our positions and could never come to Him, He came to Us! He came to save, nourish, clean, heal, and protect. He humbled Himself to take our place, to be our servant, our hero, our protector.

Who, being in the form of God, did not consider it robbery to be equal with God, but made Himself of no reputation, taking the form of a bondservant, and coming in the likeness of men. And being found in appearance as a man, He humbled Himself and became obedient to the point of death, even the death of the cross. (Phil 2:6-8)

That's love!

But God demonstrates His love for us, in that while we were still sinners, Christ died for us. (Rom 5:8)

There is one problem with the whole thing, the cell itself.

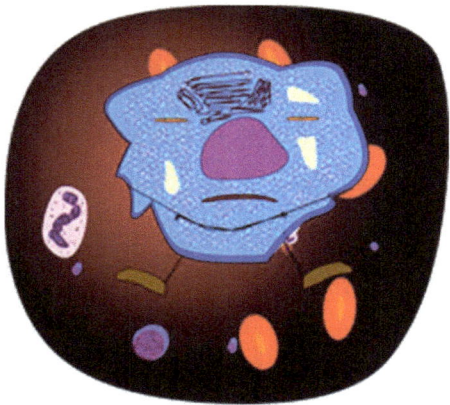

Each cell has an outer membrane, and this membrane is a barrier to all the good stuff the blood provides. Food, oxygen, and other nutrients must travel through the membrane into the cell to do any good. If the membrane does not open for these things, the cell cannot benefit. If the waste products and carbon dioxide do not travel through the membrane, the blood cannot take the waste away. The cell will eventually commit suicide, if it does not partake of all the "life that is in the blood". In the same way, if we do not open our hearts to the blood of Christ, we commit spiritual suicide because the spiritual life is in the Blood of Christ. We can only live if we accept Jesus' sacrifice for us.

> *Then Jesus said to them, "Most assuredly, I say to you, unless you eat the flesh of the Son of Man and drink His blood, you have no life in you. Whoever eats My flesh and drinks My blood has eternal life, and I will raise him up at the last day. For my flesh is food indeed and My blood is drink indeed. He who eats My flesh and drinks My blood abides in Me and I in him. (John 6:53-56)*

Don't be like the cell that refuses the life that is in the blood. Accept the gift of life that God offers and all that comes with it.

> *For God so loved the world that He gave His only begotten Son, that whoever believes in Him should not perish but have everlasting life. (John 3:16)*

You can pray:

> *I know Lord that I am a sinner, and I cannot save myself. I believe that Jesus died to pay the penalty for that sin and rose again on the third day to prepare a place for me in heaven. I now open my heart up to you and ask you*

to come in and save me. Please give me the life you offer through the shed blood of Jesus. Thank you for saving me. Amen.

www.ingramcontent.com/pod-product-compliance
Lightning Source LLC
Chambersburg PA
CBHW051336120626
46547CB00016B/2564